YOUR KNOWLEDGE HAS VALUE

Bibliographic information published by the German National Library:

The German National Library lists this publication in the National Bibliography; detailed bibliographic data are available on the Internet at http://dnb.dnb.de .

Imprint:

Copyright © 2016 GRIN Verlag, Open Publishing GmbH
Print and binding: Books on Demand GmbH, Norderstedt Germany
ISBN: 9783668275874

This book at GRIN:

http://www.grin.com/en/e-book/337777/strategies-of-efficient-management-of-water-in-agriculture

Rupesh Meena

Strategies of Efficient management of water in Agriculture

GRIN Publishing

GRIN - Your knowledge has value

Since its foundation in 1998, GRIN has specialized in publishing academic texts by students, college teachers and other academics as e-book and printed book. The website www.grin.com is an ideal platform for presenting term papers, final papers, scientific essays, dissertations and specialist books.

Visit us on the internet:

http://www.grin.com/

http://www.facebook.com/grincom

http://www.twitter.com/grin_com

Strategies of Efficient management of water in Agriculture

Introduction

Agriculture is the largest consumer of water. However, only a part of this agricultural water diverted is effectively used in the production of food or other agricultural commodities and the remaining does not reach the crop/plants because of losses through soil evaporation or infiltration, and used by weeds. Water is more crucial to Indian agriculture than is commonly perceived. India's water resources are under considerable stress as the race between food production and population growth surges relentlessly forward. With 4 per cent of the world's water resources and 18 per cent of its population, the country will be hard pressed to meet the water requirements of the various growing sectors. Continue population growth and the predicted impacts of climate change, including shifts in precipitation and glacier melt, makes the water challenge greater. It is particularly important for a large country like India, situated, as it is, in the tropical belt and experiences extreme variation in climate and rainfall across the country. Currently 65% of agriculture in India is rain dependent **(Singh and Kumar, 2009).** There are extreme variations in rainfall, the westernmost part getting less than 100 mm annually and the easternmost part receiving 100 times more. Floods and droughts can strike the country simultaneously at different places. For better use of water in agriculture in water-limited environments, efforts are needed from different research disciplines: agronomists, plant breeders, plant physiologists, plant biotechnologists, water engineers and others, to develop new approaches in water conservation. For example, is it possible to and or develop crops that require less water and maintain high yield productivity? Among different approaches water productivity or water use efficiency (WUE) is an efficient approach. A crop with high WUE should have greater yield than a crop with low WUE **(Meena *et al.*, 2013).**

Taking into consideration of the alarming fact of water crisis, there is a need to efficient management of water to conserve the precious input.

Efficient Water Management

Water is the most crucial input for agricultural production. Vagaries of monsoon and declining water-table due to its overuse resulted in shortage of freshwater supplies for agricultural use, which calls for an efficient use of this resource. Strategies for efficient management of water for agricultural use involves conservation of water, integrated water use, optimal allocation of water and enhancing water use efficiency by crops.

1. Conservation of water

In-situ conservation of water can be achieved by reduction of runoff loss and enhancement of infiltrated water and reduction of water losses through deep seepage and direct evaporation from soil. Runoff is reduced either by increasing the opportunity time or by infiltrability of soil or both. Opportunity time can be manipulated by land shaping, tillage, mechanical structures and vegetative barriers of water flow and infiltrability can be increased through suitable crop rotations, application of amendments, tillage, mulching etc. Water loss by deep seepage can be reduced by increasing soil-water storage capacity through enlarging the root zone of crops and increasing soil-water retentively. Direct evaporation from soil can be controlled with shallow tillage and mulching.

Ex-situ conservation of water can be achieved by harvesting of excess water in storage ponds for its reuse for irrigation purpose.

2. Integrated water use

Efficient utilization of water resources by the integrated use of water from different sources viz., by irrigation to supplement profile stored rainwater, conjunctive use of surface - water and groundwater, poor and good quality water and recycled (waste) water for irrigation. Supplemental irrigation for growing crops is an integrated use of rainwater stored in the profile and the irrigation water regardless of its source. Small (30-50 mm) early post-emergence irrigation stimulates root extension into deeper layers thus causing greater use of profile-stored water. So the water extraction obtained from the supplemental irrigation at crucial crop growth period is more than the proportionate increase in the level of supplemental irrigation, which is referred as priming effect of the supplemental irrigation. The priming effect varies with soil type, fertility level and amount of irrigation.

3. Optimal allocation of water

Optimal allocation of available water among the competing crops and optimum timing of application is to be decided under adequate and limited water supply situation so as to maximize economic returns from available water. Under adequate water supply situation optimal allocation involves timing of irrigations so that crop yields are maintained at their achievable potential, as per climatic conditions of the location. Under limited water supply situation irrigation water must be allocated so that periods of possible water deficits coincide with the least sensitive growth periods. Thus irrigation scheduling should be decided based on the water availability. The procedure for optimal allocation of water under limited water supply condition includes quantifying water use (ET or T) vs crop biomass relations and employment optimizing models with operational constraints. Crop simulation models can be used to schedule irrigation under different water availability condition.

4. Enhancing water-use efficiency

Causes of low water use efficiency and productivity
1. Improper land configuration
2. Wrong choices of crops and cropping systems
3. Unscientific irrigation scheduling
4. Poor crop management
5. Conveyance and application losses of water
6. Lack of moisture conservation practices
7. Low use of residual soil moisture

Improving on-farm water-use efficiency:

a. By growing less water demanding crops
b. Using water saving production technologies

Water-saving production technologies include soil and agronomic management that save water without a loss in crop yields leading to water use efficiency) viz selection of crops and cropping systems based on available water supplied and increasing seasonal evapotranspiration (ET). The later can be achieved by selection of irrigation method, irrigation scheduling, tillage, mulching, planting method and fertilization. The water utilized by crop is evaluated in terms of Water Use Efficiency.

Agronomic Practices:

i. **Selection of crops and cropping system**- Selection of crops and cropping systems for high water-use efficiency should be done on the basis of availability of water under rainfed crops, limited irrigated crops and fully irrigated crops.

Table 1: WUE of some important field crops in India

High	Medium	Low
Maize	Wheat	Green
Sorghum	Barley	Gram
Pearl millet	Oats	Pigeonpea
Finger millet		Soybean
Sugarcane		Peas

2. Tillage practices for moisture conservation: Tillage affects the WUE by modifying the hydrological properties of the soil and influencing root growth and canopy development of crops. The principal effects of tillage are the preparation of seedbed conducive to the germination of seed and growth of seedling, conservation of soil moisture in unirrigated/rained areas that infuence infiltration characteristics of the soil and providing adequate soil depth for optimum root growth, proper placement of seeds and fertilizers in the soil and inter-cultivation for weed control. Deep tillage (deep ploughing and sub-soiling) is considered beneficial to conserve the monsoon rainfall **(Meena *et al*, 2013)**. Improved puddler saves about 10-30 % water over *desi* plough. Puddling with disc plough and rotavator reduces water application by 65-85 cm compared to no puddling. Conservation tillage practice normally stores more plant available moisture than the conventional inversion tillage practices when other factors remain same. Off season tillage or summer ploughing opens the soil and improves infiltration and soil moisture regimes.

3. Mulching and crop residue management: About 69-70 per cent of the rainfall is lost through evaporation. Different types of mulches are stubble mulch, soil/dust mulch, straw mulch, plastic mulch and vertical mulching. Stubble mulching is based on stirring the soil with implements that leave considerably effective part of the vegetative material, crop residues or vegetative litter on the surface as a protection against erosion and for conserving moisture by favoring infiltration and reducing evaporation **(Rana et al, 2003).** Crop residue also improves soil tilth, adds organic matter to the soil and improves the water holding capacity of the soil. Mulching with crop residues can improve water use efficiency by 10-20% through reduced soil evaporation and increased plant transpiration. Mulching with crop residues during the summer fallow can increase soil water retention **(Feng, 1999).** Use of surplus crop residues as a mulching material in crops during summer months conserves water, reduces soil temperature and controls weeds. It improves the physical conditions of the soil by enhancing biological activity of soil fauna and add almost all major and minor nutrients, improve water infiltration in the soils, which are prone to crust and compaction problems and increases soil-water storage in the root zone, increases crop growth and yield, controls water and wind erosion of soil, helps in modification of soil temperature during summer (hot) and winter (cold) season and decreases infestation of weeds in fields.

4. **Direct Seeded Rice (DSR):** It is an efficient method of water saving in rice. Puddled transplanted rice (PuTPR) requires about 20 % more water than DSR and water requirement is more in PuTPR at establishment & vegetative phase. Several problems of PuTPR which are minimized in DSR are Puddling-ponding, frequent irrigation, cracks in puddled soil **(Jat et al., 2006).**

5. **Planting techniques/methods**: Another agronomic method for increasing water use efficiency is to follow proper planting techniques/methods. Broadbed and furrows (BBF) are formed for rainy season crops. For some crops like maize, vegetables etc., the field has to be laid out into ridges and furrows. Sugarcane is planted in the furrows or trenches. Crops like tobacco, tomato, chillies are planted with equal inter and intra-row spacing so as to facilitate two-way inter-cultivation (Singh *et al.*, 2012a).

6. **Furrow irrigated raised bed planting system (FIRBS)**: Furrow irrigated raised bed planting system (FIRBS) is a recently introduced concept of sowing crops like wheat, cotton, soybean, maize, and rice etc on raised beds with irrigation in furrow to obtain better crop performance. There is saving of about 30 per cent of irrigation water, 25 per cent less seed rate is required in this method, labour requirement is less, site specific weed management can be done, higher nutrient use efficiency, lodging of the crop is less as channels made in furrows acts as passage

for air. Bed plating facilitates irrigation before sowing and thus provides an opportunity for weed control prior to planting. In an experiment maximum chickpea grain yield was recorded under raised bed planting which was significantly higher by 16.8% and 15.9% over flat bed technique, during 2005-06 and 2006-07 (Pramanik *et al.* 2009) (Table 2).

Table 2: Yield and water use efficiency of Chickpea as influenced by plantingtechniques

Planting techniques	Grain yield (t/ha)		Water use efficiency (kg/ha-mm)	
	2005-06	2006-07	2005-06	2006-07
Flat bed	1.84	2.01	10.27	9.72
Raised bed	2.15	2.33	12.06	11.33
CD(P=0.05)	0.11	0.16		

Source: Pramanik *et al.* 2009

7. Rain Water Harvesting: Rain water harvesting will not only conserve the soil, its fertility and vegetation; but also could be utilized as supplementary irrigation that will be advantageous in enhancing total water supply available to crop plants during low rainfall period. For ex-situ rain water harvesting, farm ponds are made either by constructing an embankment across water coarse or by excavating a pit. Runoff water is collected from a catchment and stored in farm pond. The stored water is utilized for supplemental irrigation during long dry spells at critical stages of crop growth. Farm ponds hold a great promise as a life-saving device for rainfed crops in low and erratic rainfall areas **(Pandey *et al.*, 2012).**

8. Irrigation methods: Water use efficiency in crops can be increased by using water-saving irrigation methods. Recent innovations of micro-irrigation systems such as sprinkler and drip irrigation apply water without much loss, and can irrigate 1.5 to 3.0 times area compared to flooding for same amount of water. Drip irrigation is an efficient method of providing irrigation water directly into the soil at root zone of plants **(Pandey *et al.*, 2013).** Water can be applied through sprinkler, surface drip, subsurface drip, micro sprayers, etc. water is applied through continuous drop through emitters. It also improves the quality and size of the produce and thus enhances the marketable quality of the produce. The technology can be adopted for undulating terrain having shallow porous soils and water scarcity areas. Alternate furrow irrigation can save water up to 40% over the traditional flooding method.

9. Laser leveling of land: It is a precursor of resource conserving technique and a process of levelling the land surface ($\pm$ 2 cm) from its average elevation using laser equipped dragged buckets. It leveled the surface having 0 to 0.2 % slope so that there is uniform distribution of water may takes place and thus enhance water use efficiency. Advantages of laser land leveling are as follows: About 5% rise in area under cultivation due to removal of bunds and channels, saves 20-30% water due to uniform distribution, Increases resource (N and water) use efficiency, reduces cost of production and increase yield (15-20%).

10.Weed control: One of the main management means of obtaining more efficient water use is the elimination of weeds in crops. Weeds compete with crops for soil nutrients, water and light. Except in high rainfall areas the primary concern is the water factor because the water requirement of weeds compared to nutrient requirement is greater than that of crop plants.

11. Intercropping: Intercropping systems are generally recommended for rainfed crops to get stable yields. The total water used in intercropping system is almost the same as for sole crops, but yields are increased, thus water use effciency is higher than sole crops **(Singh *et al.*, 2013 d)**. It is found that good agronomic practices (not only improve better utilization of water but also proved an eco-friendly tool for sustainable management of plant diseases under changing climate scenario.

12. Fertilization: There is strong interaction between fertilizer rates and irrigation levels for crop yield and WUE. Application of nutrients facilitates root growth, which can extract soil moisture from deeper layers. Furthermore, application of fertilizers facilitates early development of canopy that covers the soil and intercepts more solar radiation and thereby reducing the evaporation

13. Varieties: Breeding and selecting crop cultivars that make more efficient use of soil and fertilizer N (including higher N fixation and N portioning) while maintaining productivity and crop quality has been a long-term goal of production agriculture. Development of nutrient efficient cultivars could help decrease fertilizer inputs and resulting nutrient losses to air and groundwater.

Mechanical Measures

1. Contour Farming: Contour farming involves ploughing, planting and weeding along the contour, i.e., across the slope rather than up and down. Contour ridges are used mainly in semi-arid areas to harvest water, and in higher rainfall areas for growing potatoes. Experiments shows that contour farming alone can reduce soil erosion by as much as 50% on moderate slopes. However, for slopes steeper than 10%, other measures should be combined with contour farming to enhance its effectiveness.

2.Terracing: Terraces are used in farming to cultivate sloped land. Graduated terrace steps are commonly used to farm on hilly or mountainous terrain. Terraced fields decrease erosion and surface runoff , and are effective for growing crops requiring much water, such as rice.

Chemical Methods

1. Anti-transparent: Anti-transparent are the materials or chemicals that are used to reduce the transpiration. These chemicals reduce the transpiration either by closing the stomata or by forming the film on the surface. The most common type of anti-transparants is of four types:

(i) Stomatal closing type: Most of the transpiration occurs through the stomata on the leaf surface. Some fungicides like phenyl mercuric acetate (PMA) and herbicides like Atrazine in low concentration serve as antitranspirants by inducing stomatal closing. **(ii) Film forming type:** Plastic and waxy material which form a thin film on the leaf surface and result into physical barrier. For example ethyl alcohol. **(iii) Reflectance type:** They are white materials which form a coating on the leaves and increase the leaf reflectance. By reflecting the radiation, vapour pressure gradient and thus reduce transpiration. Application of 5 percent kaolin spray reduces transpiration losses. Diatomaceous earth product (Celite), hydrated lime, calcium carbonate, magnesium carbonate, zincs sulphate etc. **(iv) Growth retardant:** These chemicals reduce shoot growth and increase root growth and thus enable the plants to resist drought. They may also induce stomatal closure.

Conclusion

Irrigation should be applied at optimum time, in optimum amount, with right methods to get higher water use efficiency from the water applied as well for better yield. There is an urgent need to adopt water conservation technologies to get higher returns per unit of money invested. These technologies are helpful in conserving our precious gift of nature (water) and at the same time uplift the economic standard of the farmers. Therefore, it is necessary to improve WUE in rainfed eco-systems to increase the economic crop production per unit of water and this goal cannot be achieved without implementation in field the different technologies for increasing water use efficiency in collaboration.

References

Jat ML, Chandana P, Sharma SK, Gill MA and Gupta RK. 2006. Laser Land Leveling-A Precursor Technology for Resource Conservation. Rice-Wheat Consortium Technical Bulletin Series 7, Rice-Wheat Consortium for the Indo-Gangetic Plains, New Delhi.

Meena BL, Singh AK, Phogat BS and Sharma HB. 2013. Effects of nutrient management and planting systems on root phenology and grain yield of wheat. Indian J. Agril. Sci. 83(6): 627-63 Singh A K and Kumar P. 2009. Nutrient management in rainfed dryland agro ecosystem in the impending climate change scenario. Agril. Situ. India. 66 (5): 265-270.

Rana NS, Singh AK, Kumar Sanjay and Kumar Sandeep. 2003. Effect of trash mulching and nitrogen application on growth yield and quality of sugarcane ratoon. Indian J. Agron. 48(2): 124-126.

Singh AK , Bhatt BP, Sundaram PK, Gupta AK and Singh Deepak. 2013a. Planting geometry to optimize growth and productivity faba bean (Vicia faba L.) and soil fertility. J. Environ. Biol. 34(1): 117-122.

Kaur Ramanjit, Sepat Seema, Rani Manisha and Rana K. S. 2014. Article On Techniques for Improving Crop Productivity and Water Use Efficiency. *www.popularkheti.info/ Issue-2-1/PK-2-1-8*

Feng, H. C. (1999). Effects of straw mulching on soil conditions and grain yield of winter wheat. Chinese Bulletin and Soil Science, 30, 174-175.

Anup Das, G. C. Munda, D.P.Patel. Article On Technological Options for Improving Nutrient and Water Use Efficiency. www.kiran.nic.in.

Lal Singh et al., 2014. Efcient Techniques to increase Water Use Efficiency under Rainfed Eco-systems. *Journal of Agri Search* 1(4): 193-200.

Pramanik SC, Singh NB and Singh KK. 2009. Yield, economics and water use efficiency of chickpea (*Cicer arietinum*) under various irrigation regimes on raised bed planting system. *Indian Journal of Agronomy* 54 (3): 315-318.